Easy Guide to Learning and Understanding

Molecular and Cell Biology

Contents

CELL BIOLOGY...8
AUTHORS...9
INTRODUCTION..10
SIZE OF CELLS..12
Some History of the development of understanding of the Cell 14
Related reading 17
WHAT IS A CELL?...18
WHAT IS THE DIFFERENCE BETWEEN ELEMENTS 23
WHAT IS LIVING?..27
WHAT IS INTERESTING ABOUT CELLBIOLOGY?.............................28
Summary...30
TYPES OF CELLS...31
PROKARYOTES..32
BACTERIA...33
EUKARYOTES...35
UNIQUE PROPERTIES OF PLANT CELLS 36
Chloroplasts 37
Vacuoles...39
Cell walls 40
PARTS OF THE CELL 41
MEMBRANES..42
Phospholipids..45
Cholesterol..47
Semi-permeability and Osmosis..48
Proteins and channels...50
Hydrophobicity 51
ORGANELLES..52
GENETIC MATERIAL..53
ENERGY SUPPLY (CHLOROPLASTS AND MITOCHONDRIA)................56
CELL DIVISION...61
CELL CYCLE...62
From Wikipedia 63
Overview..64

Details of mitosis 66
MEIOSIS..67
Crossover 68
MITOSIS...69
Prophase...71
Prometaphase..72
Metaphase 73
Anaphase...74
Telophase..75
Cytokinesis..76
GENES 77
EXPRESSION...78
TRANSLATION..79

Cell Biology

Size of cells

Cells are so small that even a cluster of these cells from a mouse only measures 50 microns.

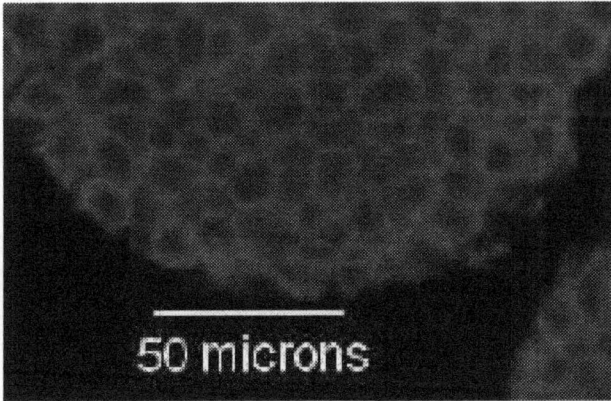

50 microns

Although it is generally the case that biological cells are too small to be seen at all with the unaided eye, there are exceptions as well as considerable range in the sizes of various cell types. Eukaryotic cells are typically 10 times the size of prokaryotic cells (these cell types are discussed in the next Chapter). Plant cells are on average some of the largest cells, probably because in many plant cells the interior is mostly a water-filled vacuole.

So, you ask, what are the relative sizes of biological molecules and cells? The following are all approximations:

0.1 nm (nanometer) diameter of a hydrogen atom

0.8 nm Amino Acid

2 nm Diameter of a DNA Alpha helix

4 nm Globular Protein

6 nm micro filaments

10 nm thickness cell membranes

11 nm Ribosome

25 nm Micro tubule

50 nm Nuclear pore

100 nm Large Virus

150-250 nm small bacteria such as Mycoplasma

200 nm Centriole

200 nm (200 to 500 nm) Lysosomes

200 nm (200 to 500 nm) Peroxisomes

400 nm giant virus Mimi virus:

1   Âµm (micrometer)

(1 -10 Âµm) the general sizes for Prokaryotes

1 Âµm Diameter of human nerve cell process

2 Âµm E.coli -a bacterium

3 Âµm Mitochondrion

5 Âµm length of chloroplast

6 Âµm (3 -10 micrometers) the Nucleus

9 Âµm Human red blood cell

10 Âµm

(10 -30 Âµm) Most Eukaryotic animal cells

(10 -100 Âµm) Most Eukaryotic plant cells

90 Âµm small Amoeba

100 Âµm Human Egg

up to 160 Âµm Megakaryocyte

up to 500 Âμm giant bacterium Thiomargarita

up to 800 Âμm large Amoeba

1mm (1 millimeter, 1/10th cm)

1 mm Diameter of the squid giant nerve cell

Some History of the Development of Understanding of the Cell

The origin of the idea that living organisms are made of cells is often traced back to observations of thin slices of cork. In 1665 the book *Micrographia: Some physiological descriptions of minute bodies made by magnifying glasses* was published by <u>Robert Hooke</u>. He wrote:

. . . I could exceedingly plainly perceive it to be all perforated and porous, much like a Honey-comb, but that the pores of it were not regular. . . . these pores, or cells, . . . were indeed the first microscopical pores I ever saw, and perhaps, that were ever seen, for I had not met with any Writer or Person, that had made any mention of them before this. . .

We now know that the "cells" Hooke observed were an indication of the cellular structure of multi-cellular organisms. During the 1670s, Antony van Leeuwenhoek used microscopes to observe sperm, red blood cells, and protozoa. While many cells are about 10 microns in diameter, some protozoa are visible to the naked eye, reaching over 1 millimeter in length. Thus, while it is true that the small size of most cells made it difficult to develop the theory that all living organisms are composed of cells, it was also difficult to recognize that living cells have certain functional components such as the nucleus and a surface membrane that allow cells to exist as the basic functional components of all living organisms. In 1833 Robert Brown published a report describing microscopic observations of plant cells in which he used first used the term "cell nucleus":

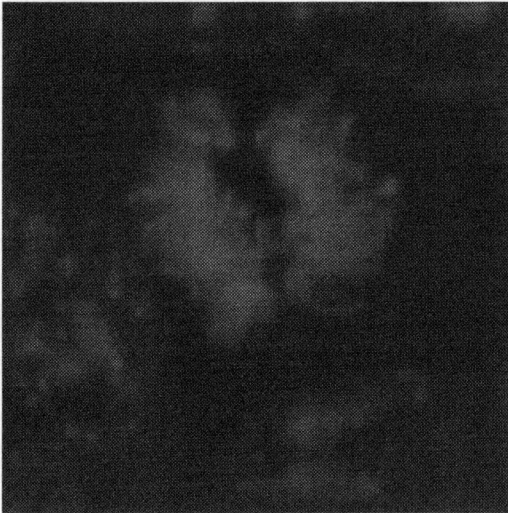

In the compressed cells of the epidermis, the nucleus is in a corresponding degree flattened; but in the internal tissue it is often nearly spherical, more or less firmly adhering to one of the walls, and projecting into the cavity of the cell.

Such observations of the microscopic cellular components of cells helped make it possible for Schleiden and Schwann to proposed a cell theory specifying that nucleated cells are key structural and functional units in plants and animals (1832-1838). However, they did not understand cell reproduction. About this time, microscopists such as the Belgian botanist Barthelemy C. Dumortier observed and reported the binary fission of cells. By 1879 the zoologist Walther Flemming was using chemical staining of "fixed" cells to allow clear visualization of <u>chromosomes</u> during cell division.

During the 1890s, Ernest Overton developed a theory of <u>lipid</u> membrane structure and function, based largely on the <u>osmotic</u> properties of cells. Visualization of lipid bilayer membranes at the surface of cells had to wait until the development of <u>electron microscopy</u>.

Related reading

What limits cell sizes?

Prokaryotes -Limited by efficient metabolism

Animal Cells (Eukaryotic) -Limited by Surface Area to Volume ratio

Plant Cells (Eukaryotic) -Have large sizes due to large central vacuole which is responsible for their growth.

What is a cell?

Cells are the fundamental building blocks of life. Cells vary to form individual "single-cell" organisms (bacteria) to "multi-cellular" structures (tissue, organs) and organisms (animals and plants).

Cells are structural units that make up plants and animals; also, there are many single-celled organisms. What all living cells have in common is that they are small 'sacks' composed mostly of water. The 'sacks' are made from a phospholipid bilayer membrane. This membrane is semi-permeable (allowing some things to pass in or out of the cell while blocking others). There exist other methods of transport across this membrane that we will get into later.

So what is in a cell? Cells are 90% fluid (called cytoplasm) which consists of free amino acids, proteins, carbohydrates, fats, and numerous other molecules. The cell environment (i.e., the contents of the cytoplasm and the nucleus, as well as the way the DNA is packed) affect gene expression/regulation, and thus are VERY important aspects of inheritance. Below are approximations of other components (each component will be discussed in more detail later):

Elements

←59% Hydrogen (H)
←24% Oxygen (O)
←11% Carbon (C)
←4% Nitrogen (N)
←2% Others - Phosphorus (P), Sulphur (S), etc.

Molecules

←50% protein
←15% nucleic acid
←15% carbohydrates
←10% lipids
←10% Other

Components of cytoplasm

Cytosol -contains mainly water and numerous molecules floating in it -all except the organelles.

←**Organelles** (which also have membranes) in 'higher' eukaryote organisms:

←**Nucleus** (in eukaryotes) - where genetic material (DNA) is located, RNA is transcribed.

←**Endoplasmic Reticulum (ER)** -Important for protein synthesis. It is a transport network for molecules destined for specific modifications and locations. There are two types:

←**Rough ER** - has ribosomes, and tends to be more in 'sheets'.

←**Smooth ER** - Does not have ribosomes and tends to be more of a tubular network.

←**Ribosomes** -half are on the Endoplasmic Reticulum, the other half are 'free' in the cytosol; this is where the RNA goes for translation into proteins.

←**Golgi Apparatus** - important for glycosylation, secretion. The Golgi Apparatus is the "UPS" of the cell. Here, proteins and other molecules are prepared for shipping outside of the cell.

←**Lysosomes** -Digestive sacks found only in animal cells; the main point of digestion.

←**Peroxisomes** -Use oxygen to carry out catabolic reactions, in both plant and animals. In this organelle, an enzyme called catalase is used to break down hydrogen peroxide into water and oxygen gas.

←**Microtubules** - made from tubulin, and make up

centrioles, cilia, etc.

←**Cytoskeleton** - Microtubules, actin and intermediate filaments.

←**Mitochondria** - convert foods into usable energy. (ATP production) A mitochondrion does this through aerobic respiration. They have 2 membranes; the inner membranes shapes differ between different types of cells, but they form projections called cristae. The mitochondrion is about the size of a bacteria, and it carries its own genetic material and ribosomes.

←**Vacuoles** - More commonly associated with plants. Plants commonly have large vacuoles.

←**Organelles found in plant cells and not in animal cells:**

←**Plastids** -membrane bound organelles used in storage and food production. These are similar to entire prokaryotic cells -for example, like mitochondria they contain their own DNA and self-replicate. They include:

←**Chloroplasts** - convert light/food into usable energy. (ATP production)

←**Leucoplasts** - store starch, proteins and lipids.

←**Chromoplasts** - contain pigments. (E.g. providing colors to flowers)

←**Cell Wall** - found in prokaryotic and plant cells; provides structural support and protection.

What is the difference between elements?

The various elements that make up the cell are:

←59% Hydrogen (H)
←24% Oxygen (O)
←11% Carbon (C)
←4% Nitrogen (N)
←2% Others - Phosphorus (P), Sulphur (S), etc.

The difference between these elements are their respective atomic weights, electrons, and in general their chemical properties. A given element can only have so many other atoms attached. For instance carbon (C) has 4 electrons in its outer shell and thus can only bind to 4 atoms; Hydrogen only has 1 electron and thus can only bind to one other atom. An example would be Methane which is CH_4. Oxygen only has 2 free electrons, and will sometimes form a double bond with a single atom, which is an 'ester' in organic chemistry (and is typically scented).

Methane Water Methanol (Methyl Alcohol)

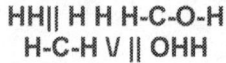

HH|| H H H-C-O-H
H-C-H V || OHH

As for the organic molecules that make up a typical cell:

←50% protein
←15% nucleic acid
←15% carbohydrates
←10% lipids
←10% Other Here is a list of Elements, symbols, weights and biological roles.
←HAHA

Element	Symbol	Atomic Weight	Biological Role
Calcium	Ca	40.1	Bone; muscle contraction, second messenger
Carbon	C	12.0	Constituent (backbone) of organic molecules
Chlorine	Cl	35.5	Digestion and photosynthesis
Copper	Cu	63.5	Part of Oxygen€"carrying pigment of mollusk blood.
Fluorine	F	19.0	For normal tooth enamel development
Hydrogen	H	1.0	Part of water and all organic molecules
Iodine	I	126.9	Part of thyroxine (a hormone)
Iron	Fe	55.8	Hemoglobin, oxygen caring pigment of many animals
Magnesium	M g	24.3	Part of chlorophyll, the photosynthetic pigment; essential to some enzymes.
Manganese	M n	54.9	Essential to some enzyme actions.
Nitrogen	N	14.0	Constituent of all proteins and nucleic acids.

Oxygen	O	16.0	Respiration; part of water; and in nearly all organic molecules.
Phosphorus	P	31.0	High energy bond in ATP.
Potassium	K	39.1	Generation of nerve impulses.
Selenium	Se	79.0	For the working of many enzymes.

What is living?

The topic "what is life?" has been one of many long discussions and the answer may depend upon your initial definitions.

Some definitions of life are:

1. The quality that distinguishes a vital and functional being from a non-living or dead body, or purely chemical matter.
2. The state of a material complex or individual characterized by the capacity to perform certain functional activities including metabolism, growth, and reproduction.
3. The sequence of physical and mental experiences that make up the existence of an individual.

Under these definitions, life may or may not include a virus that is only 'alive' if it can insert its genetic material into a living cell. To some, living systems that react to the environment, grow, improve, and reproduce are alive. A more liberal definition would include too much; a narrower one would not include all cells.

What is interesting about cell biology?

What makes Cell Biology particularly interesting is that there is so much that is not fully understood. A cell is a complex system with thousands of molecular components working together in a coordinated way to produce the phenomenon we call "life". During the 20th century, these molecular components were identified (for example, see Human Genome Project), but research continues on the details of cellular processes like the control of cell division and cell differentiation. Disruption of the normal control of cell division can cause abnormal cell behavior such as rapid tumor cell growth.

Cells have complex interactions with the surrounding environment. Whether it is the external world of a single celled organism or the other cells of a multicellular organism, a complex web of interactions is present. Study of the mechanisms by which cells respond appropriately to their environments is a major part of cell biology research and often such studies involve what is called signal transduction. For example, a hormone such as insulin interacting with the surface of a cell can result in the altered behavior of hundreds of molecular components inside the cells. This sort of complex and finely tuned cell response to an external signal is required for normal metabolism and to prevent metabolic disorders like Type II diabetes.

Most of the cells of a multi-cellular organism have the

same <u>genetic material</u> in every cell, yet, there are over 200 <u>types of cells</u> in the body that are different shapes and sizes and carry out very different functions. And ALL of these cells were developed from **1 (one)** cell (<u>zygote</u>). The study of how the many cell types develop during embryonic development (<u>Developmental Biology</u>) is a branch of Biology that is heavily dependent on the methods (such as microscopy) of Cell Biology. Much of the control of cell differentiation is at the level of the control of gene <u>transcription</u>, the control of which <u>mRNAs</u> are made. Muscle cells make muscle proteins and nerve cells make brain proteins. Geneticists, molecular biologists and cell biologists are working to discover the details of how cells specialize to accomplish hundreds of functions from <u>muscle</u> contraction to <u>memory</u> storage.

Summary

- Complexity in:
- inter-relations between cells
- signal transduction pathways inside cells
- control of cell death and cell reproduction
- control of cell differentiation
- control of cell metabolism.

Types of cells

Prokaryotes

The structures of two prokaryotic cells. The bacterium (shown at the top) is a *heterotrophs*, organisms that eat other organisms. Cyanophytes are *autotrophs*, organisms that make their food without eating other organisms.

Most of these prokaryotic cells are small, ranging from 1 to 10 microns with a diameter no greater than 1 micron. The major differences between Prokaryotic and Eukaryotic cells are that prokaryotes do not have a nucleus as a distinct organelle, and rarely have any membrane bound organelles [mitochondria, chloroplasts, endoplasmic reticulum, golgi apparatus, a cytoskeleton of microtubules and microfilaments] (the only exception may be a bacterium discovered to have vacuoles). Both types contain DNA as genetic material, have a surrounding cell membrane, have ribosomes[70 s], accomplish similar functions, and are very diverse. For instance, there are over 200 types of cells in the human body that vary greatly in size, shape, and function.

Prokaryotes are cells without a distinct nucleus. They have genetic material but that material is not enclosed within a membrane. Prokaryotes include bacteria and cyanophytes. The genetic material is a single circular DNA strand and is located within the cytoplasm. Recombination happens through transfers of plasmids (short circles of DNA that pass from one bacterium to another). Prokaryoytes do not engulf solids, nor do they have centrioles or asters. Prokaryotes have a cell wall made up of peptidoglycan.

Bacteria

Bacteria are prokaryotic, unicellular organisms. Bacteria are very small, so much so that 1 billion could fit on 1 square centimeter of space on the human gums, and 1 gram of digested food has 10 billion bacteria. Bacteria are the simplest living organisms. Previously they fell under the Kingdom Moneran, but now they fall into two different Kingdoms: Archaebacteria and Eubacteria. There are several differences between the two.

Archaebacteria have no peptidoglycan in their cellular walls. They also have odd lipids in their cell walls. Many are able to live in extreme places (like early Earth). There are 3 types of Archaebacteria. The first type is Methanogen. These use Carbon dioxide and Hydrogen to make Methane. They are found in sewage, cows, and swamps, and they do not take in oxygen. The second type is Extreme Halophile. These live in extremely salty places (i.e.: the Dead Sea and Great Salt Lake). Finally, the third type is Thermoacidophiles. These prefer extremely hot, acidic areas (i.e.: hot springs and volcanos).

Eubacteria have peptidoglycan in their cell walls, and they have no unusal lipids. They have three shapes: bacilli (hot dog shaped), cocci (ball shaped), and spirilli (spring shaped). Eubacteria can also have prefixes before

27

their names: strepto, indicating chains of the shaped bacteria, and straphylo, indicating clusters of the shaped bacteria. Eubacteria are tested in labratories for Gram stains. Gram stains will determine if antibiotics will work (Gram postive) or if they will not (Gram Negative). There are four major types of Eubacteria: Cyanobacteria (green bacteria that infest fertilizer polluted ponds and lakes and mass produce algae), Spirochetes (Gram negative bacteria on which antibiotics do not work), Gram Positive (both gram positive that are used to make yogurt, strept throat is one of these), and Proteobacteria (E-coli). Bacteria also have special structures: Plasmids (a small loop of DNA separate from the nuclear region, which is used for creating genetic variety, inserting into other organisms, and by genetic engineers) and Endospores (hard coat created by some bacteria in extreme conditions--this is why canning jars must be boiled for a long time).

Reproduction is either through binary fusion (splitting of a cell with no variety in its genes) or through several other forms that produce genetic variety: **Transformation** (taking DNA from environment and incorporating it into themselves), **Conjugation** ("sex" in which cilia hook together and the Plasmids exchange genes), and **Transduction** (viri infect the bacteria and the bacteria infects the virus with its Plasmid to move genes throughout the population).

Bacteria produce poisons that can cause sickness: exotoxins, which are given off by the Gram positive bacteria, and endotoxins, which are given off by Gram negative bacteria as they die.

Eukaryotes

Eukaryotes are cells with a distinct nucleus, a structure in which the genetic material (DNA) is contained, surrounded by a membrane much like the outer cell membrane. Eucaryotic cells are found in most algae, protozoa, and all multi-cellular organisms (plants and animals) including humans. The genetic material in the nucleus forms multiple chromosomes that are linear and complex with proteins that help the DNA 'pack', and are involved in regulation of gene expression.

The cells of higher plants differ from animal cells in that they have large vacuoles, a cell wall, chloroplasts, and a lack of lysosomes, centrioles, pseudopods, and flagella or cilia. Animal cells do not have the chloroplasts, and may or may not have cilia, pseudopods or flagella, depending on the type of cell.

Unique Properties of Plant Cells

Plant Cells have a number of important differences compared to their animal counterparts. The major ones are the Chloroplasts, Cell walls and Vacuoles. Unlike animal cells, plant cells do not have centrioles.

Chloroplasts

The chloroplasts are an organelle similar to the mitochondria in that they are self-reproducing and are the energy factories of the cell. There most of the similarities end. Chloroplasts capture light energy from the sun and convert it into ATP and sugar. In this way the cell can support itself without food.

Structure of the Chloroplasts

Vacuoles

Plants often have large structures containing water surrounded by a membrane in the center of their cells. These are vacuoles and act as a store of water and food (in seeds), a place to dump wastes and a structural support for the cell to maintain turgor. When the plant loses water, the vacuoles quickly lose their water, and when plants have a lot of water the vacuoles fill up. In mature plants there is usually one large vacuole in the center of the cell.

Cell walls

Plant cells are not flaccid like animal cells and have a rigid cell wall around them made of fibrils of cellulose embedded in a matrix of several other kinds of polymers, such as pectin and lignin. The cellulose molecules are linear and provide the perfect shape for intermolecular hydrogen bonding to produce long, stiff fibrils. It is the cell wall that is primarily responsible for ensuring the cell does not burst in hypertonic surroundings.

Parts of the cell

Membranes

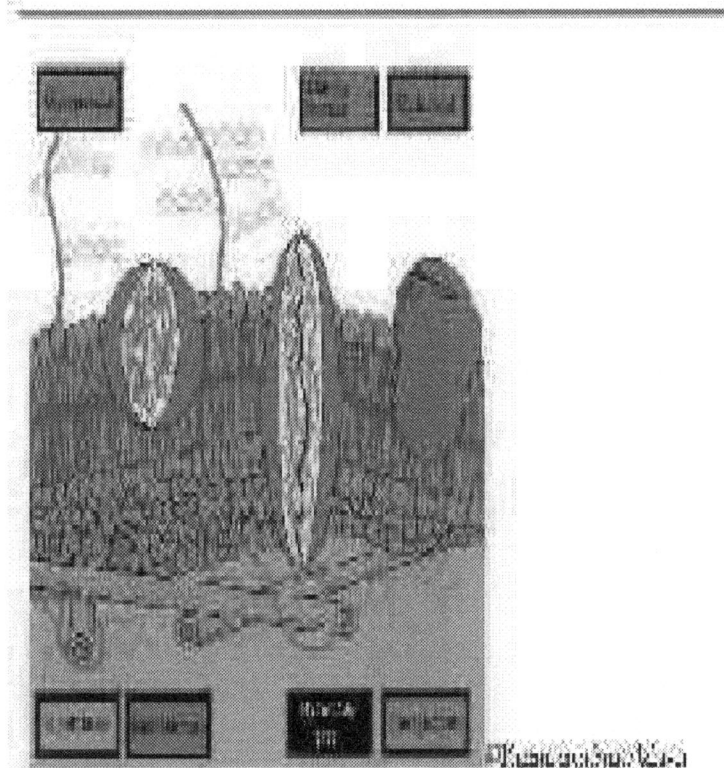

The phospholipid bilayer which the cell membrane is an example of, is composed of various cholesterol, phospholipids, glycolipids and proteins. Below is an example of a simple phospholipid bilayer.

The smaller molecules shown between the phospholids is

Cholesterol, which helps to give rigidity or stability to the membrane. The two main components of phospholipids are shown in these figures by blue circles representing the hydrophilic head groups, and by long thin lines representing the hydrophobic fatty acid tails.

Both the interior of the cell and the area surrounding the cell is made up of water or of an aqueous solution. Consequently, phospholipids orient themselves with respect to the water and with each other so that the hydrophilic ("water loving") head groups are grouped together and face the water, and so that the hydrophobic ("water fearing") tails turn away from the water and toward each other. This self-organization of phospholipids results in one of just a few easily recognizable structures. Cell membranes are constructed of a phospholipid bilayer as shown above.

Smaller structures can also form, known as 'micelles', in which there is no *inner* layer of phospholipids with their head groups oriented towards an internal aqueous space. Instead, the interior of a micell is wholly hydrophobic, filled with the fatty acid chains of the phospholipids and any other hydrophobic molecule they enclose. Micelles are not so important for the understanding of cellular structure, but are useful for demonstrating the principles of hydrophilicity and hydrophobicity, and for contrasting with lipid bilayers.

At least 10 different types of lipids can commonly be found in cell membranes. Each type of cell or organelle will have a differing percentage of each lipid, protein, and carbohydrate. The main types of lipids are:

- Cholesterol
- Glycolipids
- Phosphatidylcholine
- Sphingomyelin
- Phosphatidylethnolamine
- Phosphatydilinositol
- Phosphatidylserine
- Phosphatidylglycerol
- Diphosphatidylglycerol (Cardiolipin)
- Phosphatidic acid

Phospholipids

Phospholipids are made up of a hydrophilic head and a hydrophobic tail. The head group has a 'special' region that changes between various phospholipids. This head group will differ between cell membranes [types of cells] or different concentrations of specific 'head groups'. The fatty acid tails also differ, but there is always one saturated and one unsaturated 'leg' of the tail.

/wiki/Image:Phospholipids.png/wiki/Image:Phospholipids.png

Phospholipids are 2 fatty acids, one saturated and one

unsaturated (shown by the double bond) that are linked to a glycerol.

Cholesterol

Cholesterol is a major component of cell membranes and serves many other functions as well. Cholesterol helps to 'pack' phospholipids in the membranes, thus giving more rigidity to the membranes. In colder conditions cholesterol also serves to keep the fluidity in the cell membrane, by keeping space in between the phospholipids. Also, cholesterol serves diverse functions such as: it is converted to vitamin D (if irradiated with Ultra Violet light), modified to form steroid hormones, and is modified to bile acids to digest fats.

Semi-permeability and Osmosis

The membranes of cells are a fluid; they are semi-permeable, which means some things can pass through the membrane through osmosis or diffusion. The rate of diffusion will vary depending on its size, polarity, charge and concentration on the inside of the membrane versus the concentration on the outside of the membrane. When something is permeable, it means that something can spread throughout, like "the perfume is permeating the room". Here is a list of some molecules and how they relate to passing through the membrane without assistance, in other words, through osmosis:

Hydrophobic Molecules

- O_2 - Oxygen
- N_2 - Nitrogen
- Benzene

Small Uncharged Polar Molecules

- H_2O - Water
- Urea
- Glycerol
- CO_2 - Carbon Dioxide

Large Uncharged Polar Molecules

- Glucose
- Sucrose

Ions

- H^+ - Hydrogen ion
- Na^+ - Sodium ion
- K^+ - Potassium ion
- Ca^{2+} - Calcium ion
- Cl^- - Chloride ion

Various substances will pass through the membranes at varying rates through osmosis.

Proteins and Channels

One role of proteins in cells is the transport of molecules/ions into or out of cells. Three methods of doing this are through active, facilitated or passive transport. Other roles are in cell recognition, receptors, and cell-to-cell communication. There is more information on membrane proteins and other proteins in later sections.

Hydrophobicity

Very simplistically, hydrophilic and hydrophobic are, respectively, the like and dislike. Hydrophilic areas of a phospholipid or a protein are 'attracted' to water, and hydrophobic regions are repelled by water.

Organelles

- Nucleus: contains genetic material or DNA in the form of chromatin.
- Mitochondria: site responsible for cell's respiration. It synthesizes ATP through a protein called ATP synthase. It has an external and an internal membrane. The internal membrane is invaginated to maximize surface area to hold more ATP synthases. Has a double membrane, an outer membrane, and a folded inner membrane.
- Chloroplasts: found only in photosynthesizing cells (e.g., plants); site of photosynthesis through several photosystem proteins.
- Ribosomes: responsible for protein synthesis, it is composed of two subunits that elongate in an amino acid sequence.
- Endoplasmic Reticulum: usually it is the structure to which ribosomes are attached.
- Golgi apparatus: attaches functional groups to different biomolecules to direct them to their

respective destinations.

- Vacuole: stores water or cell wastes. Also helps plant cells retain their structure.
- Peroxisomes: performs a variety of metabolic processes and produces hydrogen peroxide as a by-product. Uses the peroxase enzyme to break down this hydrogen peroxide into water and oxygen.

For more info go to:

http://www.tvdsb.on.ca/westmin/science/sbi3a1/Cells/cells.htm

Genetic material

1. Genetic material of Prokaryotes

2. Genetic material of Eukaryotes

3. Nucleus

4. Nuclear membrane

5. Nucleolus

6. Codons

7. RNA polymerase

8. Histones

Nucleus

In cell biology, the nucleus is an organelle, found in most eukaryotic cells, which contains most of the cell's genetic material. Nuclei have two primary functions: to control chemical reactions within the cytoplasm and to store information needed for cellular division.

Nucleolus

The nucleolus is the only structure in the nucleus that is detectable in a light microscope without staining. It is the site of rRNA synthesis. (rRNA or ribosomal RNA is an essential part of ribosomes, the structures in the Cytoplasma where proteins are made from mRNA templates.) The nucleolus contains many rDNA genes from which the rRNA is made. The human genome contains 10 chromosomes with arrays of rDNA genes (two copies of chromosomes 13, 14, 15, 21 and 22) but not all of these arrays are necessarily actively used in all cells.

Energy Supply (Chloroplasts and Mitochondria)

Chloroplasts are the organelles that incorporate energy into storage, while mitochondria are the ones that release the energy from the stores.

1Glycolysis
2Krebs cycle
3Electron transport

Glycolysis

Glycolysis is the initial step in breaking down the food that we consume for energy. Glycolysis can take two paths: either fermentation or the grooming step. In respiration, the Glucose molecules that have been formed from digestion are broken down into Glyceraldehyde-3-Phosphate, taking in 2ATP molecules for energy, and giving out NADH. This Glyceraldehyde-3-Phosphate molecule is converted into Pyruvate, giving out 4ATP molecules and NADH molecules. The Pyruvate is converted into Acetyl coenzyme A, taking in coenzyme A, and releasing several compounds such as CO_2 and NADH; but this is part of the next step in respiration, called the Link Reaction.

Fermentation happens when the organism is in an anaerobic state (lacking oxygen, like during periods of high physical activity or simply lack of air). The large Glycolysis molecule is taken and split into 2 Pyruvic Acid molecules. In animals, this Pyruvic acid is turned into lactic acid; this is why our muscles burn during workouts. In plants, the Pyruvic Acid is broken down into Ethyl Alcohol and CO_2, where fermentation comes from.

If the organism has oxygen, then the 2 Pyruvic acid (A 3 Carbon molecule) is mixed with CoA and produces 2 Acetal CoA(A 2 Carbon molecule), 2NADH's, and $2CO_2$'s. This is also our net gain for glycolysis/grooming step.

Krebs cycle

The Krebs Cycle is also known as the Citric Acid Cycle or Tricarboxylic Acid (TCA) cycle and occurs within the Matrix of the Mitocondria.

Acetyl-Coenzyme A produced from the Link Reaction is synthesized into **Citric Acid** by the action of citrate synthase and its combination with **Oxaloacetate**. The citric acid, known as **Citrate** in its ionized form, is then converted into **Isocitrate** by aconitase. This is then converted to **Alpha-Ketoglutarate** by citrate dehydrogenase in an oxidative decarboxylation reaction, in which one molecule of CO_2 is given off and one molecule of NAD+ is oxidized to NADH. Another oxidative decarboxylation converts the alpha-ketoglutarate to **Succinyl Coenzyme A** with the addition of CoA-SH, and again the loss of a CO_2 and the oxidation of NAD+ is seen. Succinyl CoA is then converted into **Succinate** by the action of Succinyl CoA Synthetase, the addition of water, the loss of CoA-SH and the substrate level phosphorylation of one molecule of ADP to ATP. Succinate is then converted to **Fumarate** by Succinate Dehydrogenase, a dehydrogenation reaction which results in the reduction of FADH+ to FADH2. Fumarate is, in turn, hydrated by the addition of one water molecule by Fumarase to **Malate,** which is finally converted back to **Oxaloacetate** by Malate Dehydrogenase in a dehydrogenation reaction that oxidizes one NAD+ to NADH. The cycle can then begin again.

There are many products of, and substrates for the Krebs cycle: for instance, amino acids can be metabolized to or catabolised from **Acetyl CoA, Alpha-Ketoglutarate, Succinyl CoA, Malate** and **Oxaloacetate**, depending which specific amino acid. **Pyruvate** can be re-synthesized by Pyruvate carboxylase from **Oxaloacetate**, which can either replenish the Krebs Cycle or make glucose by **Gluconeogenesis**. **Citrate** can be used for fatty acid (fat) or cholesterol synthesis. **Succinyl CoA** can be used to make Haem for Red Blood Cells.

NADH and FADH2 are then used in the **electron transport chain** in the inner mitochondrial membrane.

Net gain for the Krebs Cycle: 2 Carbon Dioxide, 1 ATP, 3 NADH, and 1 FADH2 **per molecule of Acetyl CoA**

Electron transport

This stage involves using the NADH molecules produced previously, along with electrons from reduced electron carriers, to form molecules of ATP.

A significant part of all cells is the electron transport chain. In plant cells, energy is used from the sun to start chemical reactions that create ATP.

Cell división

Cell cycle

The normal cell cycle consists of 3 major stages. The first is Interphase, during which the cell lives and grows larger. The second is <u>Mitosis,</u> when the cell divides. The final one is Cytokinesis, which is when the two daughter cells complete their separation.

The **cell cycle** is the cycle of a biological cell, consisting of repeated mitotic cell division and interphase (the growth phase).

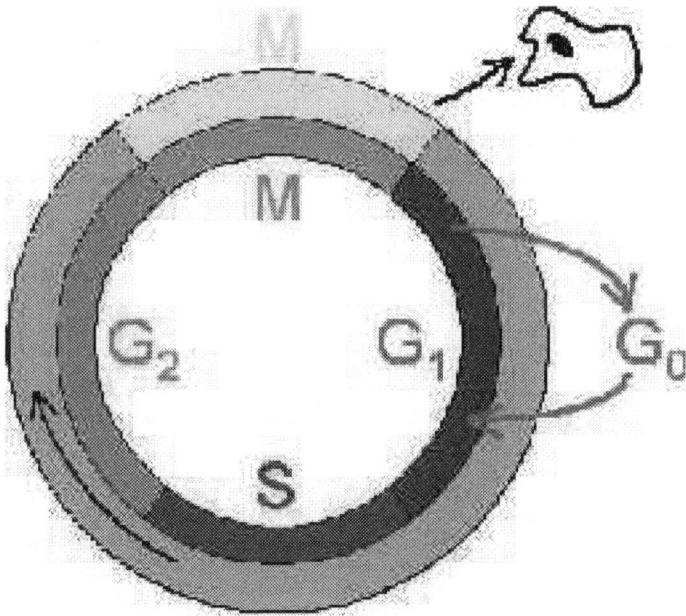

Schematic of the cell cycle. I = Interphase, M = Mitosis. The duration of mitosis in relation to the other phases has been exaggerated in this diagram.

The cell cycle consists of:

- **G1 phase**, the first growth phase
- **S phase**, during which the DNA is replicated, where S stands for the Synthesis of DNA.
- **G2 phase** is the second growth phase, also the preparation phase for the
- **M phase** or <u>mitosis</u> and <u>cytokinesis,</u> the actual <u>division</u> of the cell into two daughter cells

The cell cycle stops at several checkpoints and can only proceed if certain conditions are met, for example, if the cell has reached a certain diameter. Some cells, such as <u>neurons</u>, never divide once they become locked in a G_0 phase

Details of Mitosis

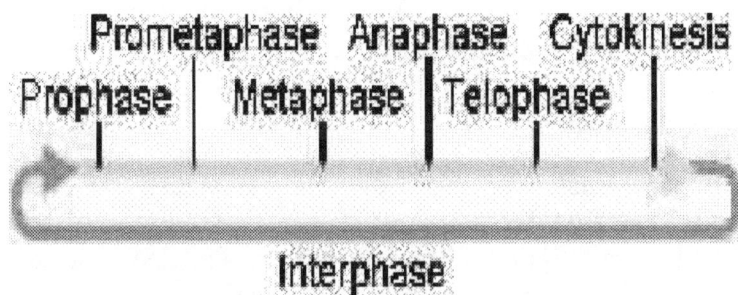

Schematic of interphase (brown) and mitosis (yellow).

Meiosis

Meiosis is a special type of cell division that is designed to produce gametes. Like normal cell division, the cell will be double diploid and have a pair of each chromosome.

Meiosis consists of 2 cell divisions, and results in four cells. The first division is when genetic crossover occurs and the traits on the chromosomes are shuffled. The cell will perform a normal prophase, then enter metaphase during which it begins the crossover, then proceed normally through anaphase and telophase.

The first division produces two normal diploid cells, however the process is not complete. The cell will prepare for another division and enter a second prophase. During the second metaphase, the chromosome pairs are separated so that each new cell will get half the normal genes. The cell division will continue thorough anaphase and telophase, and the nuclei will reassemble. The result of the divisions will be 4 haploid gamete cells.

Crossover

Crossover is the process by which two chromosomes paired up during prophase I of meiosis exchange a distal portion of their DNA. Crossover occurs when two chromosomes, normally two homologous instances of the same chromosome, break and connect to each other's ends. If they break at the same locus, this merely results in an exchange of genes. This is the normal way in which crossover occurs. If they break at different loci, the result is a duplication of genes on one chromosome and a deletion on the other. If they break on opposite sides of the centromere, this results in one chromosome being lost during cell division.

Any pair of homologous chromosomes may be expected to cross over three or four times during meiosis. This aids evolution by increasing independent assortment, and reducing the genetic linkage between genes on the same chromosome.

Mitosis

Mitosis is the normal type of cell division. Before the cells can divide, the chromosomes will have duplicated and the cell will have twice the normal set of genes.

The first step of cell division is **prophase**, during which the nucleus dissolves and the chromosomes begin migration to the midline of the cell. (Some biology textbooks insert a phase called "prometaphase" at this point.) The second step, known as **metaphase**, occurs when all the chromosomes are aligned in pairs along the midline of the cell. As the cell enters **anaphase**, the chromatids, which form the chromosomes, will separate and drift toward opposite poles of the cell. As the separated chromatids - now termed chromosomes - reach the poles, the cell will enter **telophase** and nuclei will start to reform. The process of mitosis ends after the nuclei have reformed and the cell membrane begins to separate the cell into two daughter cells, during **cytokinesis**.

From Wikipedia:

In biology, Mitosis is the process of chromosome segregation and nuclear division that follows replication of the genetic material in eukaryotic cells. This process assures that each daughter nucleus receives a complete copy of the organism's genome. In most eukaryotes, mitosis is accompanied with cell division or cytokinesis, but there are many exceptions, for instance among the fungi. There is another process called meiosis, in which the daughter nuclei receive half the chromosomes of the parent, which is involved in gamete formation and other similar processes.

Mitosis is divided into several stages, with the remainder of the cell's growth cycle considered interphase. Properly speaking, a typical cell cycle involves a series of stages: G1, the first growth phase; S, where the genetic material is duplicated; G2, the second growth phase; and M, where the nucleus divides through mitosis. Mitosis is divided into prophase, prometaphase, metaphase, anaphase, and telophase.

The whole procedure is very similar among most eukaryotes, with only minor variations. As prokaryotes lack a nucleus and only have a single chromosome with no centromere, they cannot be properly said to undergo mitosis.

Prophase

The genetic material (DNA), which normally exists in the form of chromatin, condenses into a highly ordered structure called a chromosome. Since the genetic material has been duplicated, there are two identical copies of each chromosome in the cell. Identical chromosomes (called sister chromosomes) are attached to each other at a DNA element present on every chromosome called the centromere. When chromosomes are paired up and attached, each individual chromosome in the pair is called a chromatid, while the whole unit (confusingly) is called a chromosome. Just to be even more confusing, when the chromatids separate, they are no longer called chromatids, but are called chromosomes again. The task of mitosis is to assure that one copy of each sister chromatid - and only one copy - goes to each daughter cell after cell division.

The other important piece of hardware in mitosis is the centriole, which serves as a sort of anchor. During prophase, the two centrioles - which replicate independently of mitosis - begin recruiting microtubules (which may be thought of as cellular ropes or poles) and forming a mitotic spindle between them. By increasing the length of the spindle (growing the microtubules), the centrioles push apart to opposite ends of the cell nucleus. It should be noted that many eukaryotes, for instance plants, lack centrioles although the basic process is still similar.

Prometaphase

Some biology texts do not include this phase, considering it a part of prophase. In this phase, the nuclear membrane dissolves in some eukaryotes, reforming later once mitosis is complete. This is called open mitosis, found in most multi-cellular forms. Many protists undergo closed mitosis, in which the nuclear membrane persists throughout.

Now kinetochores begin to form at the centromeres. This is a complex structure that may be thought of as an 'eyelet' for the microtubule 'rope' - it is the attaching point by which chromosomes may be secured. The kinetochore is an enormously complex structure that is not yet fully understood. Two kinetochores form on each chromosome - one for each chromatid.

When the spindle grows to sufficient length, the microtubules begin searching for kinetochores to attach to.

Metaphase

As microtubules find and attach to kinetochores, they begin to line up in the middle of the cell. Proper segragation requires that every kinetochore be attached to a microtubule before separation begins. It is thought that unattached kinetochores control this process by generating a signal - the mitotic spindle checkpoint - that tells the cell to wait before proceeding to anaphase. There are many theories as to how this is accomplished, some of them involving the generation of tension when both microtubules are attached to the kinetochore.

When chromosomes are bivalently attached - when both kinetochores are attached to microtubules emanating from each centriole - they line up in the middle of the spindle, forming what is called the metaphase plate. This does not occur in every organism - in some cases chromosomes move back and forth between the centrioles randomly, only roughly lining up along the midline.

Anaphase

Anaphase is the stage of meiosis or mitosis when chromosomes separate and move to opposite poles of the cell (opposite ends of the nuclear spindle). Centromeres are broken and chromatids rip apart.

When every kinetochore is attached to a microtubule and the chromosomes have lined up along the middle of the spindle, the cell proceeds to anaphase. This is divided into two phases. First, the proteins that bind the sister chromatids together are cloven, allowing them to separate. They are pulled apart by the microtubules, towards the respective centrioles to which they are attached. Next, the spindle axis elongates, driving the centrioles (and the set of chromosomes to which they are attached) apart to opposite ends of the cell. These two stages are sometimes called 'early' and 'late' anaphase.

At the end of anaphase, the cell has succeeded in separating identical copies of the genetic material into two distinct populations.

Telophase

Now the nuclear membrane reforms around the genetic material and the chromosomes are unfolded back into chromatin. This is often followed by cytokinesis or cleavage, where the cellular membrane pinches off between the two newly separated nuclei, to form two new daughter cells.

Cytokinesis

Cytokinesis refers to the d of a eukaryotic cell. Cytokinesis generally follows the replication of the cell's chromosomes, usually mitotically, but sometimes meiotically. Except for some special cases, the amount of cytoplasm in each daughter cell is the same. In animal cells, the cell membrane forms a cleavage furrow and pinches apart like a balloon. In plant cells, a cell plate forms, which becomes the new cell wall separating the daughters. Various patterns occur in other groups.

Genes

Expression

Gene expression is the first stage of a process that decodes what the DNA holds in a cell. It is the expression of a gene that gives rise to a protein.

How Does Gene Expression Occur?

It starts of with transcription that gives rise to the middlemen, namely the RNA. The RNA relay information from the chromosomal DNA to the cytoplasm where the machinery for protein synthesis resides.

Translation occurs following transcription wherein the protein synthesis machinery gets into action and uses its tools to read out the message that the RNA holds. The details of this process are indeed very complex and will probably be dealt with in an advanced write-up.

Translation

The Translation Phase of Genetic Expression is divided into 2 Steps: Transcription and Translation. During Transcription, RNA Polymerase unzips the two halves of the DNA where it needs to transcript. Then, free RNA bases attach to the DNA bases with the Polymerase, starting at the promoter and ending at the Termination signal. From this, the RNA can become mRNA, rRNA, or tRNA. The mRNA is a ribbon-like strand that takes the genetic information from the nucleus of the cell to the ribosome. rRNA forms a globular ball that attaches to the rough E.R. to help make ribosomes. Finally, the tRNA forms a hair-shaped landing base that reads the genetic information to make proteins.

Translation happens when mRNA is pulled through a ribosome and tRNA reads the RNA bases on the mRNA to make anti-codons of 3 bases, and brings amino acids to form the protein. This starts with the condon AUG and ends at UAG. When done, the protein forms the correct shape and does the task it was created for. This brings the genetic code from the nucleus - which it never leaves - to the cytoplasm of the cell, where proteins are produced to upkeep the body.

We Have Book Recommendations For You

The Strangest Secret by Earl Nightingale (Audio CD – January, 2006)

Acres of Diamonds [MP3 AUDIO] [UNABRIDGED] (Audio CD) by Russell H. Conwell

Automatic Wealth: The Secrets of the Millionaire Mind-- Including: Acres of Diamonds, As a Man Thinketh, I Dare you!, The Science of Getting Rich, The Way to Wealth, and Think and Grow Rich [UNABRIDGED] by Napoleon Hill, et al (CD-ROM)

Think and Grow Rich [MP3 AUDIO] [UNABRIDGED] by Napoleon Hill, Jason McCoy (Narrator)   (Audio CD - January 30, 2006)

As a Man Thinketh [UNABRIDGED] by James Allen, Jason McCoy (Narrator) (Audio CD)

65

Your Invisible Power: How to Attain Your Desires by Letting Your Subconscious Mind Work for You [MP3 AUDIO] [UNABRIDGED]
by Genevieve Behrend, Jason McCoy (Narrator) (Audio CD)

Thought Vibration or the Law of Attraction in the Thought World [MP3 AUDIO] [UNABRIDGED]

by William Walker Atkinson, Jason McCoy (Narrator) (Audio CD - July 1, 2005)

The Law of Success Volume I: The Principles of Self-Mastery by Napoleon Hill (Audio CD - Feb 21, 2006)

The Law of Success, Volume I: The Principles of Self-Mastery (Law of Success, Vol 1) (The Law of Success) by Napoleon Hill (Paperback - June 20, 2006)

The Law of Success , Volume II & III: A Definite Chief Aim & Self-Confidence by Napoleon Hill (Paperback - June 20, 2006)

Thought Vibration or the Law of Attraction in the Thought World & Your Invisible Power (Paperback)

Automatic Wealth, The Secrets of the Millionaire Mind - Including: As a Man Thinketh, The Science of Getting Rich, The Way to Wealth and Think and Grow Rich (Paperback)

The Bestsellers in this Book give sound advice about money or how to obtain it. Just shoot for the stars, stay focused on your dreams, and they will come true. There is nothing that we can imagine that we can't do. So what are we waiting for? Let's begin the journey of self-fulfillment.

4 Bestsellers in 1 Book:

As a Man Thinketh by James Allen

The Science of Getting Rich by Wallace D. Wattles

The Way to Wealth by Benjamin Franklin

Think and Grow Rich by Napoleon Hill

Get Published!

BN Publishing helped authors publish more titles. So whether you're writing a romance novel, historical fiction, mystery, action and suspense, poetry, children's or any other genre, we can help you reach your publishing goals.

Paperback

Reach 20,000 retail accounts in the U.S. (including chains, independents, specialty stories, and libraries).

Including:

www.amazon.com

www.amazon.co.uk

www.amazon.ca

www.bn.com

www.powells.com

www.ebay.com

and more...

Your book will be included in a physical catalog that will go out to over 20,000 retail stores.

When your title is entered into our library it will automatically appear in the bookstore and library databases.

Our United States and United Kingdom based sales teams works with publisher clients based throughout the world who want to print books in the United States and United Kingdom, or reach the North American, UK and wider European markets through our broad distribution channel partners.

If we decide to publish it:

We will send you 2 free copies of the finished book; we will give you 10% royalty of the selling price of each book copy sold (selling price = the price the book is sold by BN Publishing to wholesalers or other resellers); and if you wish to have more copies of your book, we will sell you the book for two thirds of the list price.

Please send us more information about your book to info@bnpublishing.com

www.bnpublishing.com